Jagadish Kumar Mogaraju

GIS and Geostatistics. GIS applications in Groundwater studies

Groundwater Quality at Y.S.R. Kadapa District, Andhra Pradesh, India

GRIN Verlag

Bibliografische Information der Deutschen Nationalbibliothek:

Die Deutsche Bibliothek verzeichnet diese Publikation in der Deutschen National-
bibliografie; detaillierte bibliografische Daten sind im Internet über http://dnb.d-
nb.de/ abrufbar.

Imprint:

Copyright © 2013 GRIN Verlag GmbH
Druck und Bindung: Books on Demand GmbH, Norderstedt Germany
ISBN: 978-3-656-58941-9

GIS and Geostatistics

GIS applications in Groundwater studies

Jagadish Kumar Mogaraju

Case study 1: Determination of an optimal interpolation technique by using ordinary kriging to represent spatial distribution of Groundwater Quality Indices at YSR District, Andhra Pradesh, India

Abstract- This work emphasized on Geostatistical tools which relied on ordinary kriging technique to analyze and represent groundwater quality in Y.S.R district, Andhra Pradesh, India whilst maintaining prediction accuracy. The efficiency of the semivariograms of kriging interpolation technique was compared to best predict key groundwater quality indices such as pH,TH,SAR,Na^+,Mg^{2+},Ca^{2+},Cl^-,HCO_3,TDS,EC and Groundwater levels in the study area. The exploratory data analysis along with cross-validation was performed. The best semivariogram model selection was made using ArcGIS 9.3. Prediction maps were prepared after systematic analysis based on Root Mean Square Error (RMSE) values.

Keywords- Geostatistical analysis, Nugget, Ordinary Kriging, Prediction maps, Sill, Spatial dependence.

1. Introduction

Groundwater is a commodity which is intended to be used judiciously whilst protecting its serenity and sanctity in terms of quality and quantity. Ubiquitous utilization in sectors such as industrial, municipal, commercial, agricultural and residential makes groundwater contaminated and converting it as a vulnerable entity. Population growth is in the forefront to create enhanced water demand due to everlasting shortage of surface water and overweening industrialization. Geographic Information Systems (GIS) initiated a beneficial symbiotic relationship with environmental concerns and natural resources in recent times. Vacuity in between GIS analysis and geostatistics is effectively bridged by ArcGIS Geostatistical analyst module. Several studies were attempted employing interpolation techniques devoid of Geostatistical tool and along with it. Hu et al (2005) conducted a study in which spatial variability existed in groundwater quality in Central North China was effectively determined using ordinary kriging. Zhu et al (1996) prepared a spatial distribution map of radon by employing GIS techniques and kriging in Belgium. D'Agostino et al (1998) compared ordinary kriging and co-kriging techniques whilst studying the spatial distribution of nitrate concentrations in an aquifer of central portion of Italy . Istok and Cooper (1998) showed that spherical model was the best fitted model for experimenting variograms of sulphate, Chloride and EC. The aims of this investigation are to provide an overview of current groundwater quality for key parameters such as pH, TH, Sodium Absorption Ratio (SAR), Na^+, Mg^{2+},Ca^{2+},Cl^-, HCO_3,Total Dissolved Solids (TDS), Electrical Conductivity (EC), Groundwater level (GWL) and to represent the spatial distribution of key parameters of the study area using Geostatistical tools and GIS techniques.

2. Materials and Methods

The study area is one of the districts in Rayalaseema region extending from 78^0 to 79^0 longitudes to and 14^0 to 15^0 latitudes. The study is flourished with forest and water resources all over. The area is blessed with divine deities of Lord Venkateswara and Lord Rama who are popular in all parts of the world.

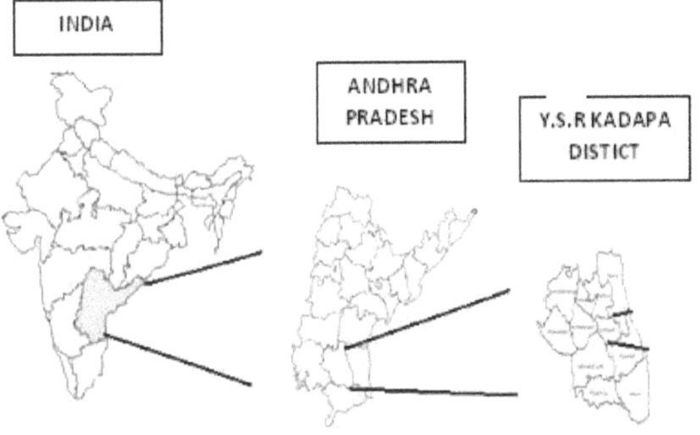

The water samples were collected from bore wells all over the district during January 2010 and January 2011.The toposheets of YSR district was obtained at 1:50000 scales and were digitized with UTM coordinate system through on-screen digitization. The software's used for the entire study were ArcGIS Geostatistical analyst and ArcGIS along with kriging methods for interpolation. Kriging technique is the optimal linear prediction of processes which are spatially linked. This technique is used in environmental monitoring, hydrology, geology and allied branches for interpolation of spatial data. The statistical and mathematical properties of the sample points are taken into consideration by Geostatistical techniques for interpolation such as kriging. The spatial configuration and autocorrelation of the measured points around the location can be quantified using Geostatistical techniques. As stated by Nas B (2009), it's been an interesting assumption that in kriging the data is from a stochastic stationary process while other methods involve normally distributed data. The kriging interpolation technique was performed in two systematic tasks starting with quantification of spatial structure followed by production of prediction maps. This technique utilizes fitted models derived from variography to predict unknown values of a specified sample location. Several variants such as ordinary kriging, simple kriging, universal kriging, co-kriging, block kriging and disjunctive kriging are in general use for various studies. Among all the variants, it was felt that the most reliable method is ordinary kriging after a systematic statistical comparative analysis. The semivariogram plot was obtained and the values were fitted into the models such as Circular (C), Spherical (S), Tetraspherical (T), Pentaspherical (P), Exponential (E), Gaussian (G), Rationale quartile (R), Hole effect (H), K-Bessel (K), J-Bessel (J) and Stable (S). Due to prediction accuracy and simplicity when compared with other kriging methods, ordinary kriging method was used. In the current study, semivariogram models are tested against each data set parameter. The model which provided best predictions was done by performing cross validation. Spatial dependence among the groundwater quality indices were also tested by calculating nugget and sill percentages for each model along with the parameter dataset. According to Taghizadeh et al (2008), If (Nugget/Sill) % is <25% then can be inferred that the variable has a strong spatial dependence, if the value ranges between 25-75% then it can be inferred that moderate spatial dependence exists. If the calculated value is greater than 75% then the variable has weak spatial dependence.

3. Results and Discussion

The samples were collected from 36 wells in the study area. The summary statistics can be viewed in Table 1. It was observed that pH, TH, SAR, Na, Mg, Ca, Cl, HCO3 ,TDS, Electrical conductivity and groundwater level never showed normal distribution. Since the data for each parameter is asymmetrical, log transformation was applied in order to make distribution to normal. Semivariogram models were tested for each dataset. The prediction accuracies for each model was analyzed through cross validation and can be viewed from Table 2 through Table 11. The maximum and minimum values of pH was 7 and 8.35 respectively and the best fit is the circular model with a Standardized Mean Error (SME) of 0.01234 and it is close to zero with a Root Mean Square standardized of 0.7942. If the Root mean square standardized error value is close to the average estimated prediction standard error, then it is said to be appropriate. For pH and TDS it is observed that Circular model is the best fit. Likewise all the parameters were tested which yielded the best fitted models. It was observed that for TH, SAR, Na, Mg, Ca, HCO3, GWL J-Bessel model was the best fit. For Chloride and Electrical conductivity, the best fitted model was the Hole effect. Spatial dependence was also calculated and tabulated in Table 12. After calculating the best model for each parameter dataset, prediction maps were prepared which can be seen from figure 1 through figure 11.

4. Conclusions

Groundwater is quintessential entity in the study area which comprises urban and peri-urban localities. Since the standard of living is low most of the population depends on groundwater wells since two to three decades. The objective of the present study is evaluating the groundwater quality of Y.S.R district, Andhra Pradesh, India and produce prediction maps. The spatial distribution of the key parameters such as pH, total hardness, SAR, sodium, Magnesium, Calcium, Chloride, Bicarbonates, Total dissolved solids, Electrical conductivity and groundwater level were successfully done via Geostatistical and GIS techniques. After checking the data for normal distribution with the help of histograms and QQ plots, it was decided that ordinary kriging is the appropriate method along with semivariograms will be the best to prepare prediction maps. Cross validation was performed to analyze the accuracy. It is sensed that there is an immediate need to monitor and mitigate groundwater quality by the local authorities. It is also advised to take into consideration the prediction maps before drilling new wells and any other establishments where groundwater is linked.

5. Illustrations

GWQI	Min	Max	Mean	Median	Std Dev	Skewness	Kurtosis
pH	7	8.35	7.6409	7.62	0.37965	0.040014	2.2757
TH	11.6	350.84	144.42	116.47	91.148	0.72165	2.466
SAR	0.86	11.286	4.3992	3.55	2.6242	1.0739	3.6639
Na	25	500	195.06	192.5	116.69	1.0746	4.0982
Mg	10	206.84	70.812	46.95	53.822	0.97703	3.1425
Ca	1.6	152	73.675	57	45.563	0.41012	1.822
Cl	40	815	270.16	184	214.46	1.1649	3.5036
HCO3	100	769	348.16	335.5	179.97	0.1423	2.7789
TDS	8.51	4179	644.55	438.5	943.09	2.1174	7.6713
EC	7.89	3900	1609.7	1346.5	1186	0.54956	2.362

| GWL | 1.15 | 24.16 | 7.0744 | 5.66 | 5.4487 | 1.3009 | 4.821 |

Table 1. Summary Statistics

Model	Mean	RMS	Average SE	Mean std	RMSS
C	0.005246	0.404	0.5344	0.01234	0.7942
S	0.009022	0.3976	0.5158	0.01988	0.7928
T	0.00825	0.3991	0.5185	0.01805	0.7964
P	0.008218	0.3993	0.5203	0.1788	0.795
E	0.008006	0.3975	0.5259	-0.01729	0.7773
G	0.01193	0.3999	0.5161	-0.0244	0.8066
R	0.01014	0.395	0.5201	-0.02109	0.7753
H	0.01028	0.4019	0.4818	-0.02385	0.8782
K	0.0115	0.3999	0.5179	-0.02339	0.8044
J	0.1268	0.4001	0.5074	-0.0263	0.8183
S	0.01196	0.3999	0.5158	-0.02446	0.8073

Table 2. Selection of best fit for pH through cross validation.

Model	Mean	RMS	Avg SE	Mean Std	RMSS
C	3.231	86.175	134.3	-0.06274	0.8378
S	4.066	87.17	131.4	-0.05081	0.8131
T	3.55	87.99	130.8	-0.05768	0.8079
P	4.266	87.75	131.7	-0.04584	0.7927
E	4.042	91.63	135.6	-0.05147	0.8105
G	3.46	86.23	136.3	-0.05872	0.8287
R	3.309	86.4	136.2	-0.03463	0.7443
H	3.54	79.28	119.5	-0.02723	0.8007
K	3.219	86.49	133.4	-0.6032	0.8252
J	4.27	79.13	112.9	-0.02	0.8482
S	3.46	86.23	136.23	-0.05872	0.8287

Table 3. Selection of best fit for TH through cross validation

Model	Mean	RMS	Avg SE	Mean Std	RMSS
C	0.09869	2.635	3.02	0.002244	0.9333
S	0.1061	2.645	3.006	0.003051	0.94
T	0.108	2.65	2.996	0.003535	0.9425
P	0.1096	2.653	2.922	0.003954	0.9438
E	0.1105	2.684	3.03	0.003777	0.9356
G	0.1298	2.624	3.033	0.009264	0.9306
R	0.112	2.659	3.023	0.005378	0.9307
H	0.1153	2.626	2.908	0.007462	0.9702
K	0.1276	2.626	3.03	0.1057	0.9311
J	0.1181	2.631	2.858	0.008504	0.9884
S	0.1297	2.625	3,.033	0.0092	0.9305

Table 4. Selection of best fit for SAR through cross validation

Model	Mean	RMS	Avg SE	Mean Std	RMSS
C	6.013	121.3	149.7	0.02058	0.8712
S	6.18	121.2	152.9	0.02311	0.8518
T	6.319	121.4	153.7	0.0237	0.8476
P	6.695	122.1	150.7	0.2646	0.8667
E	7.763	124.6	154	0.03074	0.8476

	4.17	115.4	161.4	0.01166	0.793
G					
R	8.643	123.4	150.1	0.03585	0.8584
H	5.582	116.3	138.3	0.02407	0.9186
K	4.149	115.4	165.1	0.0111	0.7756
J	5.73	116.3	133.4	0.02612	0.9507
S	4.17	115.4	161.4	0.01167	0.763

Table 5. Selection of best fit for Na through cross validation

Model	Mean	RMS	Avg SE	Mean Std	RMS Std
C	2.889	46.64	85.66	-0.02243	0.6895
S	3.189	46.79	80.47	-0.2539	0.7203
T	3.529	47.21	79.93	-0.2274	0.7332
P	3.579	47.41	79.79	-0.01708	0.7287
E	3.451	49.07	82.12	-0.02062	0.736
G	4.754	46.73	87.26	-0.003003	0.7312
R	3.129	48.71	82.06	-0.01221	0.6829
H	3.356	45.29	75.59	-0.01112	0.7054
K	4.292	47.74	83.67	0.003933	0.721
J	3.283	45.38	73.29	-0.01092	0.7242
S	3.854	48.3	84.79	-0.01087	0.7364

Table 6. Selection of best fit for Mg through cross validation

Model	Mean	RMS	Avg SE	Mean Std	RMSS
C	5.859	46.94	100.7	-0.003542	0.5907
S	6.035	47.67	102.6	-0.004476	0.5509
T	6.247	47.72	103.3	0.002434	0.5509
P	6.383	47.77	103.8	0.006864	0.5442
E	5.744	50.21	104.3	-0.006142	0.5729
G	5.765	46.24	102.1	-0.003972	0.5674
R	5.666	46.68	107.7	0.008616	0.5051
H	6.391	44.32	91.34	0.03606	0.5927
K	5.725	46.53	102.8	-0.006016	0.5662
J	6.849	45.18	85.18	0.04442	0.6524
S	5.765	46.24	102.1	-0.003971	0.5673

Table 7. Selection of best fit for Ca through cross validation

Model	Mean	RMS	Avg SE	Mean Std	RMS Std
C	-1.47	236.1	332.6	-0.03527	0.7944
S	-0.5128	237	313/1	-0.03517	0.8466
T	0.2318	237.7	301.4	-0.03454	0.8813
P	0.8179	238.1	293.7	-0.0338	0.9059
E	1.149	238.4	294.7	-0.0331	0.8984
G	-1.272	232.1	357.3	-0.03204	0.7341
R	2.357	237.9	283.8	-0.02974	0.9317
H	3.125	236.7	274.8	-0.03003	0.9734
K	-0.5708	233	345.4	-0.0316	0.7621
J	2.579	236.3	287	-0.02977	0.9301
S	-1.272	232.1	357.3	-0.03204	0.7341

Table 8. Selection of best fit for Cl through cross validation

Model	Mean	RMS	Avg SE	Mean Std	RMSS
C	-3.667	186.8	198.3	-0.01701	0.9473
S	-3.73	187.2	198.4	-0.01721	0.9483
T	-4.265	187.2	198.5	-0.01972	0.9477
P	-4.865	187.2	198.4	-0.02327	0.9479
E	-5.054	190.7	199	-0.02242	0.9604
G	-3.172	187.5	198.6	-0.01704	0.9489
R	-6.105	186.6	198.3	-0.02964	0.9454
H	-12.06	188.9	196.8	-0.06078	0.9672
K	-3.75	187.6	198.7	-0.01716	0.9491
J	-12.2	159.7	193.6	-0.05117	0.8436
S	-3.711	187.5	198.6	-0.01704	0.9489

Table 9. Selection of best fit for bicarbonate through cross validation

Model	Mean	RMS	Avg SE	Mean Std	RMSS
C	42.4	967.4	1094	0.03201	0.957
P	57.36	949.1	1123	0.04529	0.8839
T	65.77	948.6	1123	0.05258	0.8826
P	63.3	948.1	1124	0.05092	0.8809
E	59.28	959.3	1130	0.04639	0.8901
G	52.22	941.17	1118	0.0432	0.881
R	69.78	959	1152	0.05691	0.8481
H	65.75	986.1	1649	0.05822	1.008
K	58.36	939.9	1120	0.04587	0.8788
J	68.25	943.7	1144	0.5705	0.8401
S	55.23	941.6	1118	0.04321	0.8809

Table 10. Selection of best fit for TDS through cross validation

GWQI	Model	Nugget(Co)	Sill(Co)	(Co/Co+C)%	Dependence
pH	C	0.00079984	0.00478074	16%	Strong
TH	J	0.25975	0.56721	45%	Moderate
SAR	J	0.29414	0.395	74%	Moderate
Na	J	0.29531	0.52429	56%	Moderate
Mg	J	0.37942	0.78749	48%	Moderate
Ca	J	0.36924	0.84794	43%	Moderate
Cl	H	0.45011	0.79816	56%	Moderate
HCO3	J	8153.8	34481.8	23%	Strong
TDS	C	255190	1410490	18%	Strong
EC	H	1334700	1420956	93%	Weak
GWL	J	27.981	29.885	93%	Weak

Table 11. Spatial dependence using nugget and sill

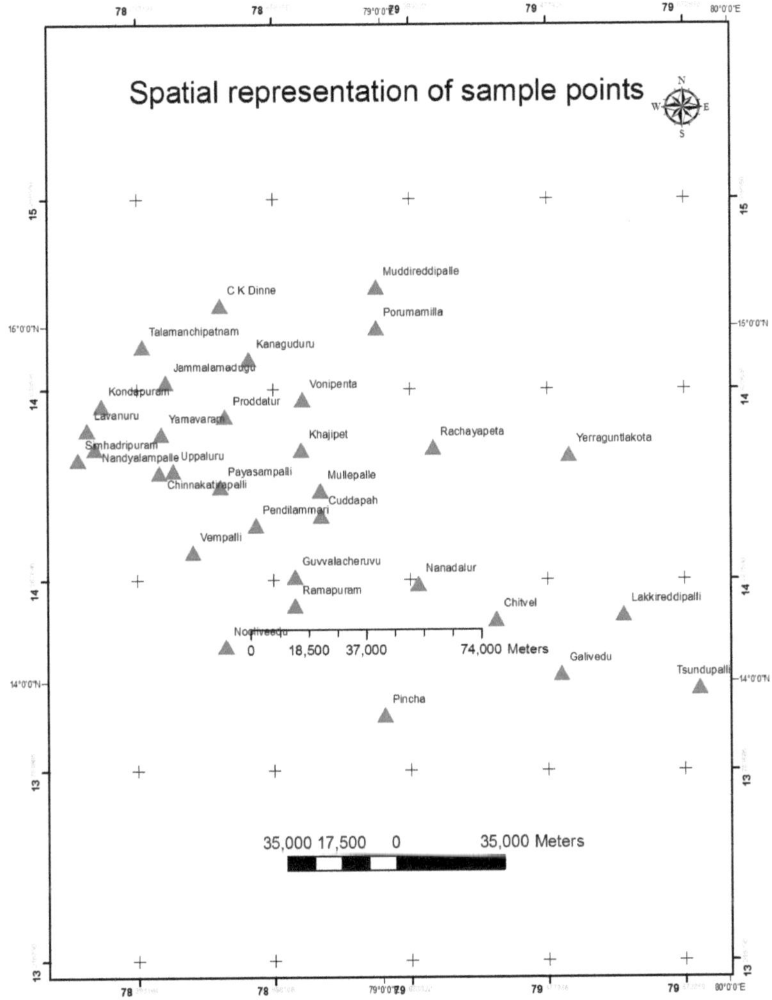

Figure 1. Spatial representation of sample points

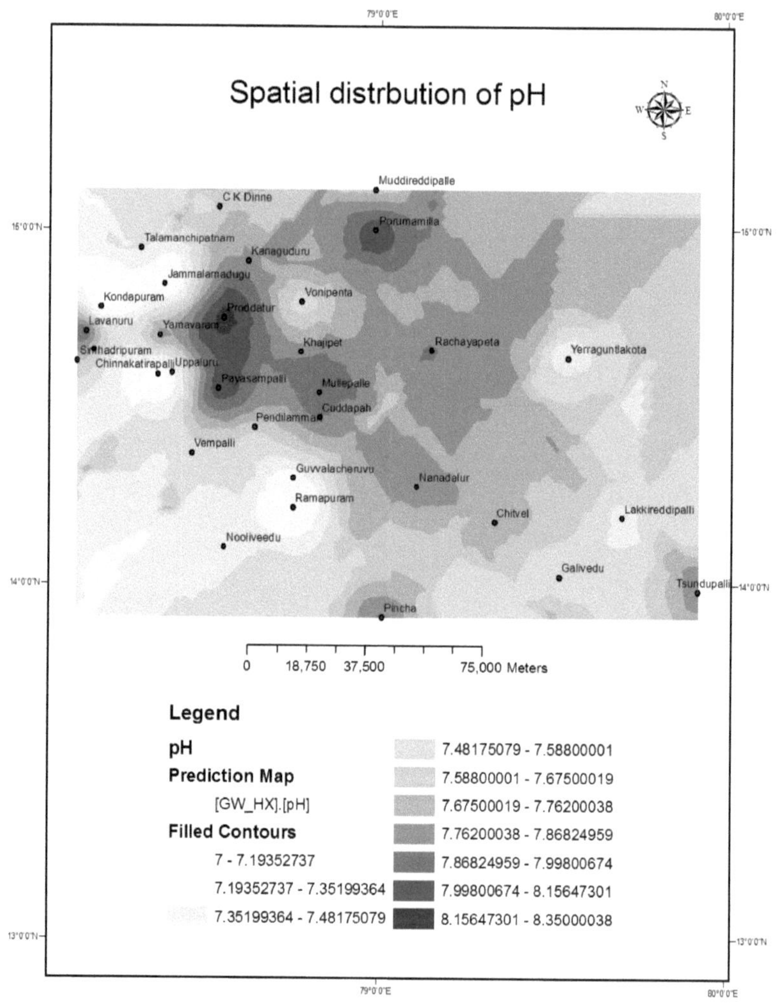

Figure 2. Spatial distribution of pH

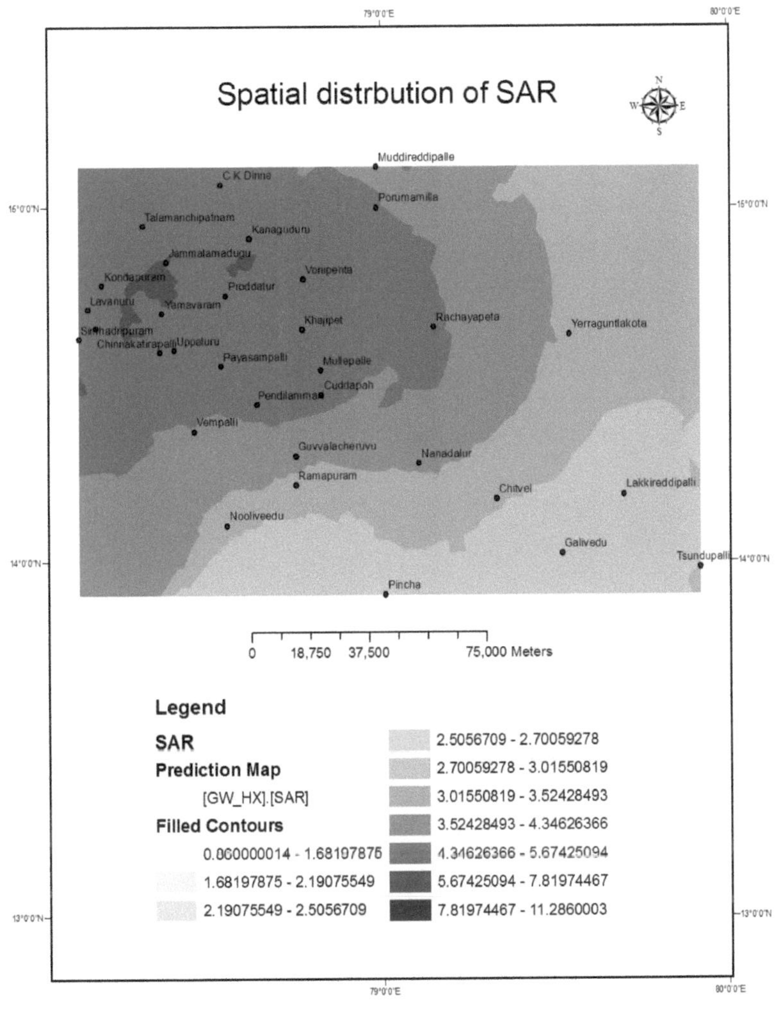

Figure 3. Spatial distribution of SAR

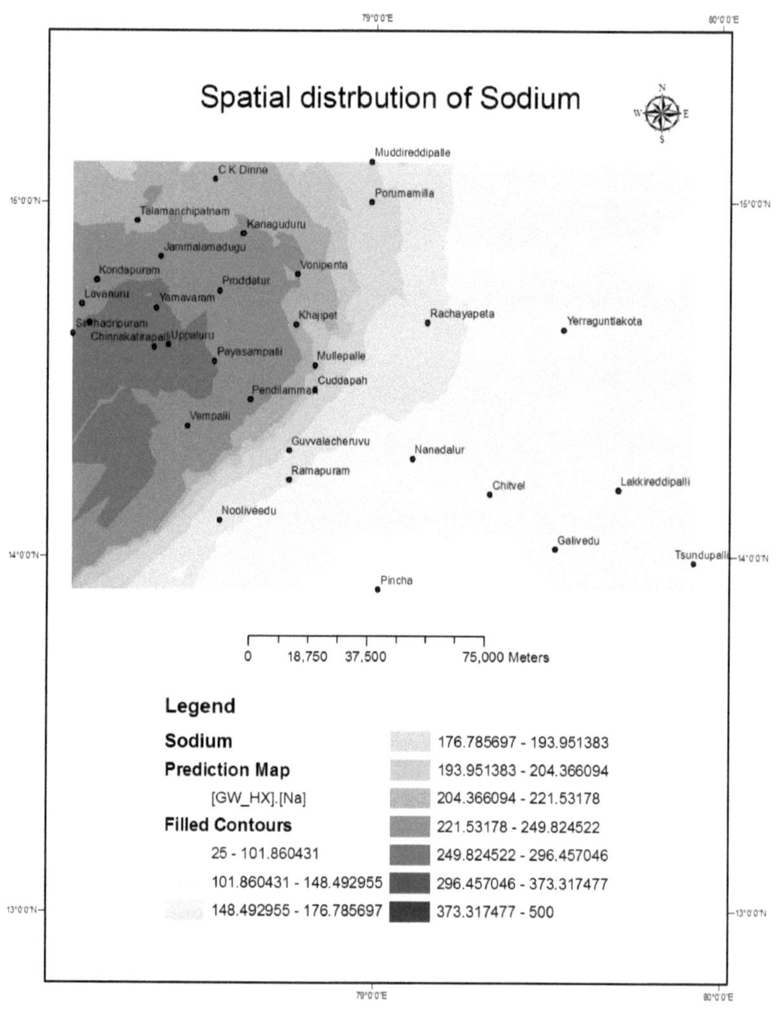

Figure 4. Spatial distribution of Sodium

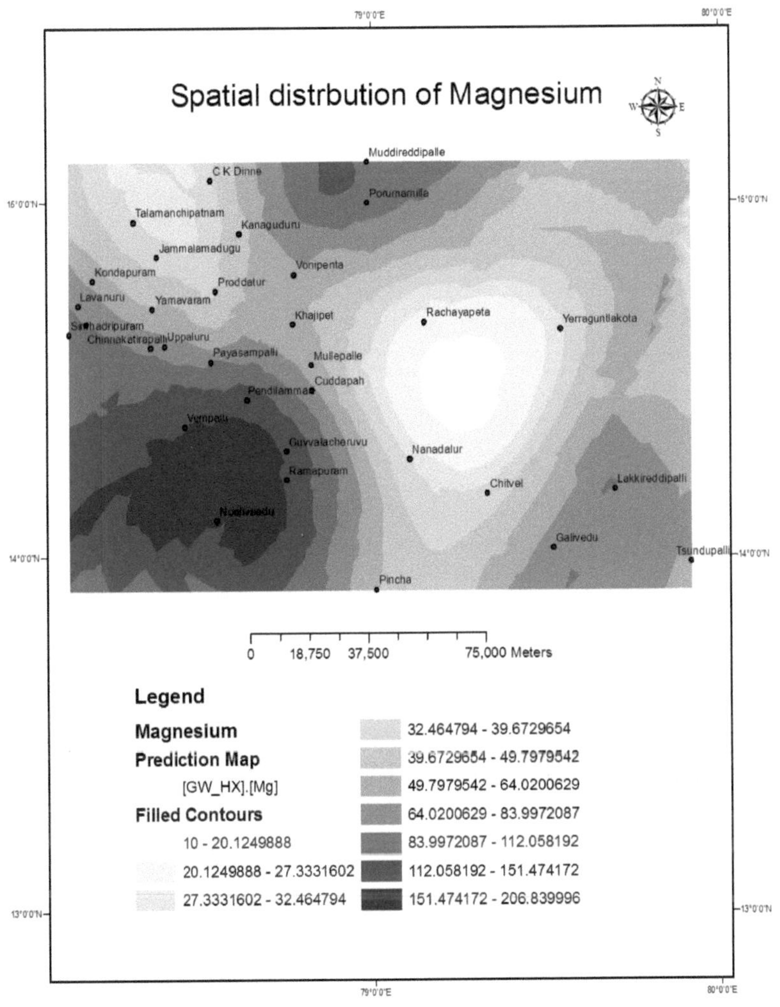

Figure 5. Spatial distribution of Magnesium

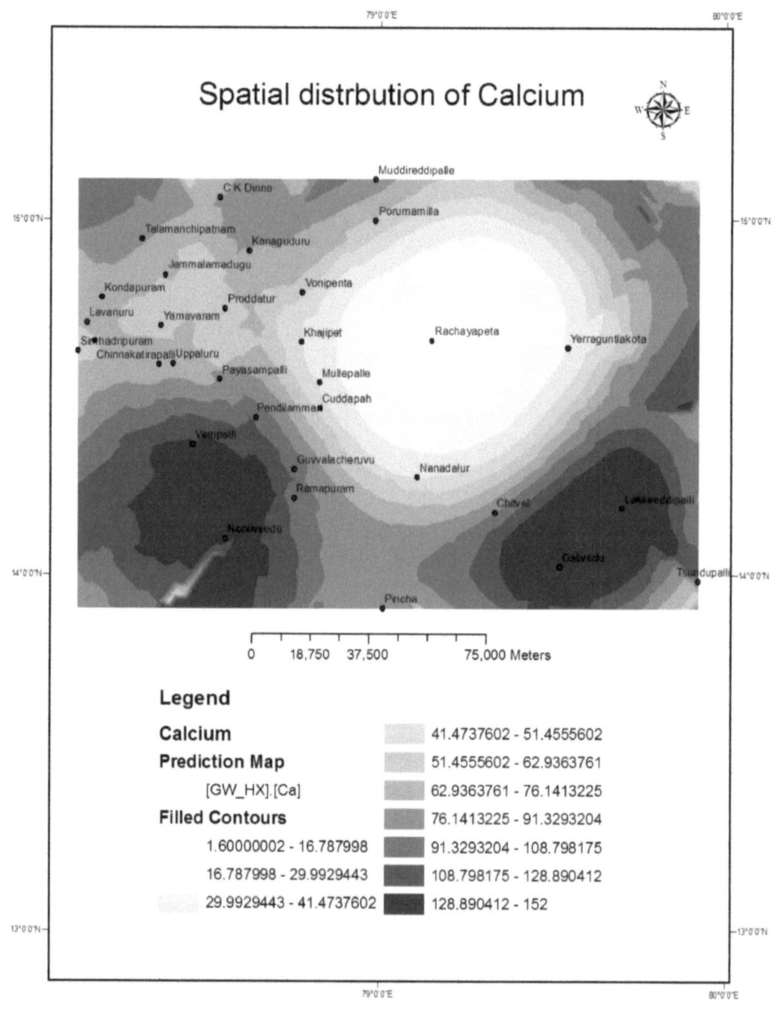

Figure 6. Spatial distribution of Calcium

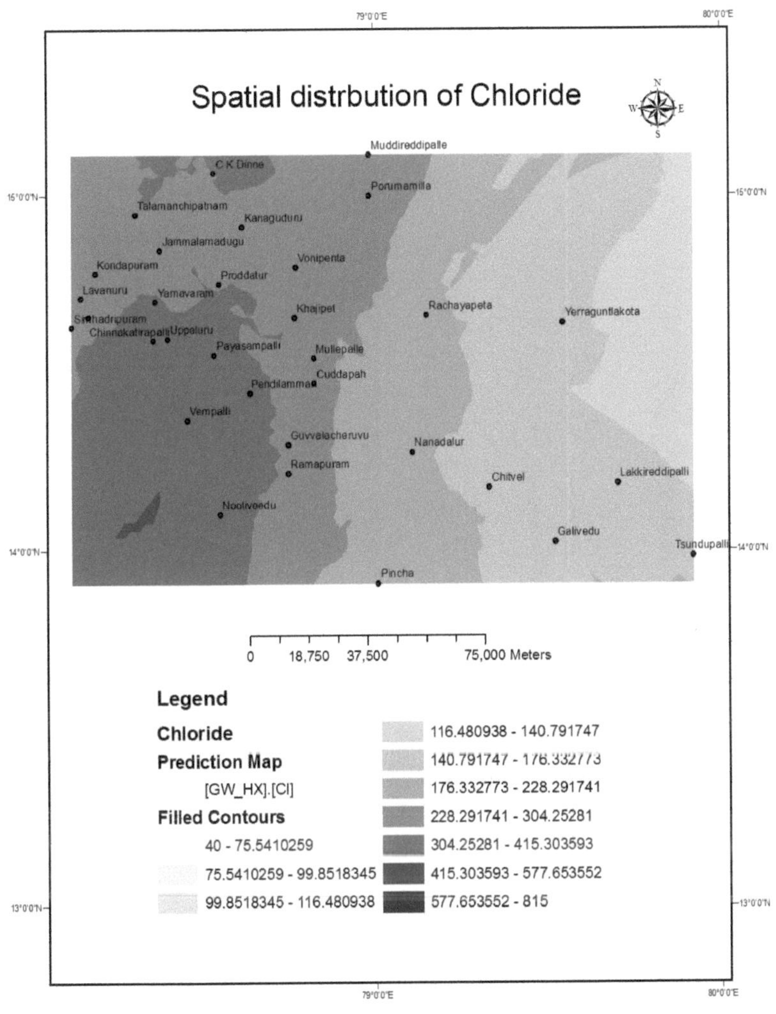

Figure 7. Spatial distribution of Chloride

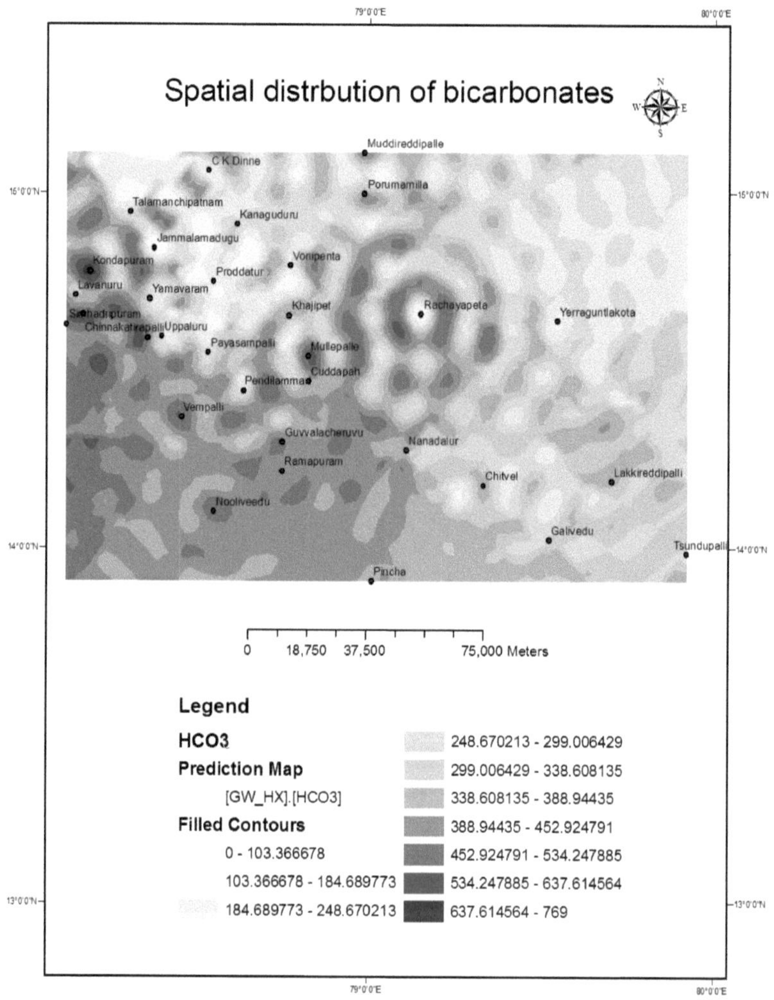

Figure 8. Spatial distribution of bicarbonates

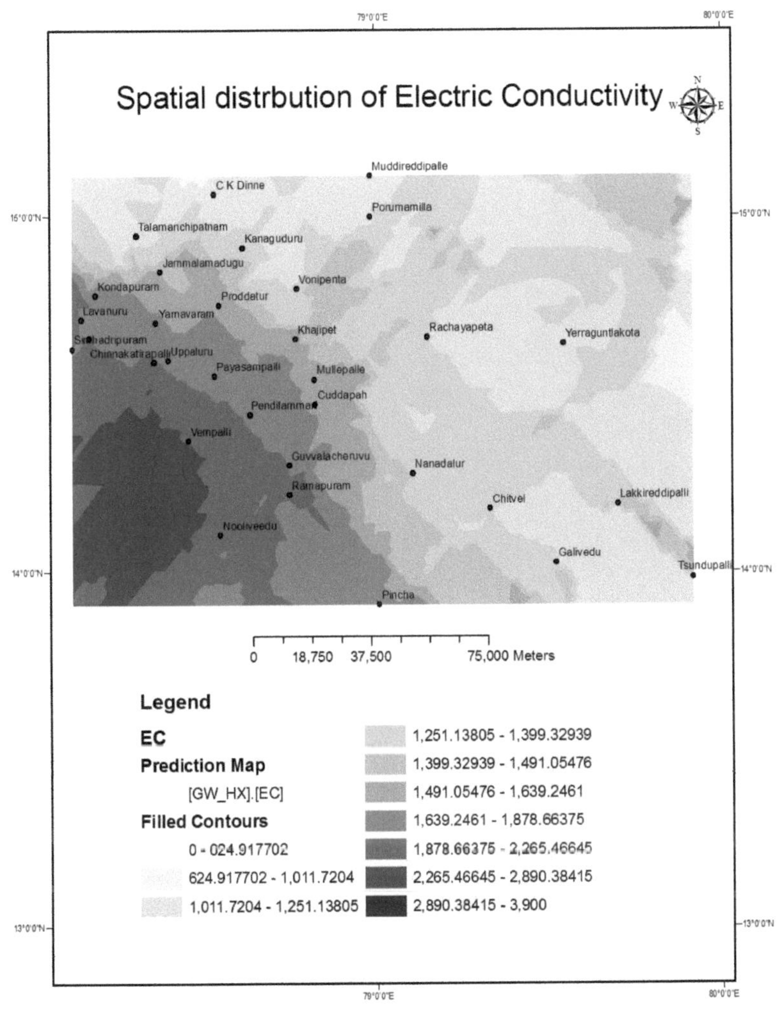

Figure 9. Spatial distribution of Electric conductivity

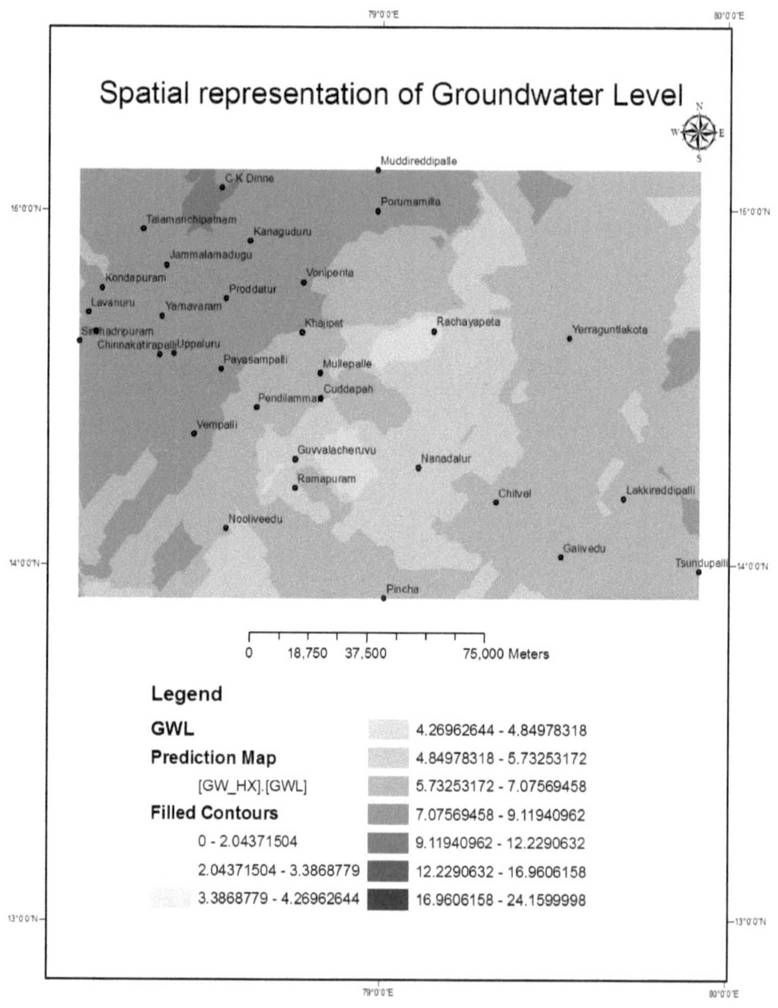

Figure 10. Spatial representation of Groundwater level

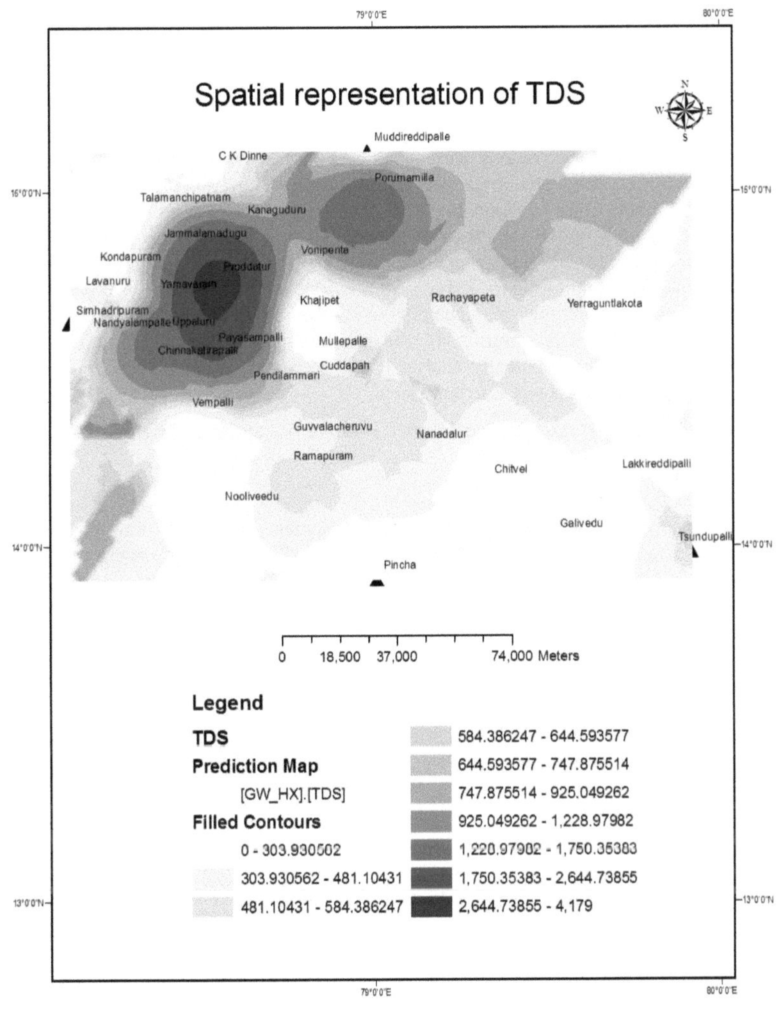

Figure 11. Spatial representation of TDS

Case study 2: Determination of an optimal interpolation technique to represent the spatial distribution of groundwater quality at urban and peri-urban areas of Proddatur, Y.S.R district, Andhra Pradesh, India

Abstract-The preliminary aim of the present study is to analyze the spatial variability of ground water quality parameters in Proddatur, India. The spatial variability and the GIS maps can be used for management of ground water quality and effective reduction of pollution related hazards can be achieved.Geostatistical analysis is a quintessential commodity to understand the ground water quality parameters and their physico chemical characteristics. The present study was conducted to analyse and assess the groundwater quality from samples collected from 43 bore holes in Proddatur, YSR district of Andhra Pradesh. Physico chemical analysis was conducted on the samples and spatial statistical analysis was performed on the data procured. Geostatistical techniques such as Inverse Distance Weighting (IDW), Radial Basis Function (RBF) and Ordinary Kriging were employed to generate surface maps after selecting the best fit among the three techniques. The key physico chemical parameters such as pH, Total Hardness (TH), Calcium (Ca^{2+}), Magnesium (Mg^{2+}), Chloride (Cl^-), Fluoride (F^-) and Alkalinity were systematically analysed and statistical treatment was conducted on them. A multivariate analysis was also conducted.

Keywords: Ground water quality, Geostatistics, Multivariate analysis, Kriging, IDW, Spline.

1. Introduction

The ground water quality is quintessential for the human growth and welfare. Since the rehabilitation of ground water is difficult if it gets polluted, the effective management of this key resource is inevitable. It is obvious that in current scenario the demand of water for industrial and non industrial needs is increasing in a geometric progression. The spatio-temporal behavior and dynamics of the ground water should be studied in order to maintain an effective management exercise. Stringent necessity to study and evaluate the ground water quality which was polluted due to point and non point sources is essential. In this part of research, the concentration of pollutants existing in unmeasured locations was studied using geostatistical methods. The main objective of the present research was to assess the ground water quality using geostatistical tools. This study was initiated to provide prediction maps and to enlighten the local authorities about the affected areas and susceptible areas which are in the line of deterioration in terms of ground water quality. The Geostatistical methods were employed to model correlation among spatial structures via variogram to determine spatial correlation measurements. Since global statistics incurred pivotal prominence over local statistics, simulation over a stochastic scale is an essential commodity in geospatial analysis. In the sub-surface hydrology studies, errors are to be avoided and enhanced rate of calculation accuracy is expected, so the Geostatistical tools serve as a handy tool to achieve these tasks. Traditional statistical sampling techniques are devoid of spatial connectivity. In contrary to the classical scenario, Geostatistics takes into account sample values in addition to their location and hence allowing the user to simultaneously study locations and values of the samples collected. The spatial structure is determined by the relating spatial properties such as distance and direction, so as to achieve a meaningful analysis.

2. Study area

The present study area is located southwest of Y.S.R district, India. The area spreads through $78^0 33'E$ to $78^0 55'E$ longitudes and $14^0 44'N$ to $14^0 73'N$ latitudes. This area falls in No-57J/6 toposheet provided by Survey of India. This area has an elevation averaged to 155 meters as per surveys conducted in the recent past. Needless to say this area is flourished with commercial acquaintances and it's considered as second Bombay by the native public. The groundwater resources in this area is vital for urban and peri urban activities prevailing in current situation.

LOCATION MAP OF STUDY AREA

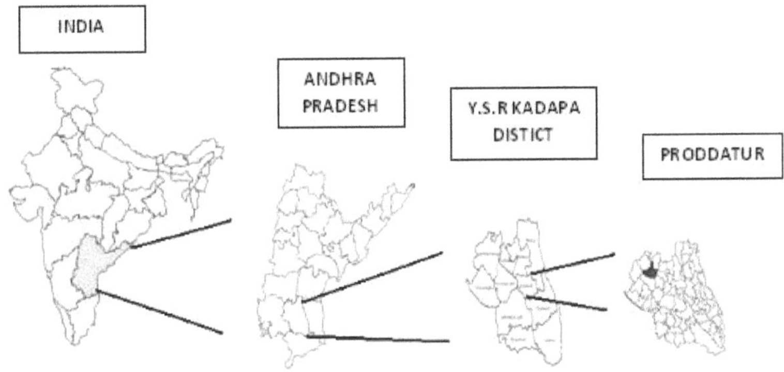

Figure 1: Study area

3. Methodology

The spatial structure is identified and modeled in the process of geostatistical prediction. The variogram is employed to study homogeneity and spatial continuity of the interested variable. The best fitted variogram decides various stages of this process. The spatial prediction methods include Inverse distance weighting (IDW), Radial basis function (Spline) and Kriging. The distribution and magnitude was evaluated whilst calibrating the data. The present work focused on the hydro geochemical parameters of the local aquifer. Factorial variables are established by employing Principal component analysis and multivariate analysis in order to summarize physico chemical data obtained from the area of interest. Root mean square error was taken into consideration whilst evaluating the performances of the selected models by cross validation mode. The lesser the RMSE values the more the predictions are. The multivariate analysis which includes factor analysis was used in order to establish interrelationship within the data. The major part of the current work is based on two software's namely ARCINFO 9.3 banked with geostatistics module and XLSTAT.

4. Results and discussion

Statistical treatment was employed to produce summary statistics for the ground water quality parameters. The summary statistics were analyzed in non-transformation mode and transformation (log) mode. The summary statistics was represented in table 2 and table 3. The factor analysis was employed to reveal the principal factors which explain the correlations among multiple outcomes. The principal factors are the reduced set of the variables which are generated using a classical method called as principal component analysis. PCA was conducted among 43 bore wells and it's followed by factor analysis. The variance of the loadings within each factor can be achieved by varimax rotation. The correlation matrix was established using the variables. The best fitted interpolation technique among the three for each variable was evaluated based on the RMSE values.

.The factor analysis was employed to reveal the principal factors which explain the correlations among multiple outcomes. The principal factors are the reduced set of the variables which are generated using a classical method called as principal component analysis. PCA was conducted among 43 bore wells and it's followed by factor analysis. The variance of the loadings within each factor can be achieved by varimax rotation. The correlation matrix was established using the variables. The best fitted interpolation technique among the three for each variable was evaluated based on the RMSE values. GIS maps were produced after systematic analysis based on geostatistics.

Table 1: Result of chemical and statistical analysis of physico-chemical parameters of ground water samples collected from the study area.

VILLAGES	PH	TH	Ca	ALKALINITY	Cl	F	Mg
Chowdur	7.90	1692.00	520.00	310.00	2420.00	1.50	284.70
Kothapeta	7.40	320.00	108.00	200.00	280.00	1.10	51.50
Shankarapuram	8.00	352.00	132.00	240.00	580.00	1.60	53.40
Chotapalli	8.00	420.00	140.00	380.00	1180.00	1.00	68.00
Dorasanipalli	7.80	320.00	124.00	680.00	1040.00	1.50	47.60
Ramapuram	8.00	308.00	104.00	400.00	420.00	1.00	49.50
Gopavaram	7.90	300.00	104.00	420.00	580.00	1.60	47.60
Bagath Singh Colony	8.10	400.00	160.00	320.00	320.00	1.10	58.30
Virat Nagar	7.90	300.00	100.00	360.00	180.00	0.90	48.60
Linga reddy Kottalu	7.90	592.00	196.00	520.00	236.00	1.50	96.20
Housing Board Colony	7.90	400.00	140.00	440.00	580.00	1.60	63.10
Rajiv Nagar	8.00	308.00	112.00	340.00	540.00	1.40	47.60
Dwaraka Nagar	7.70	600.00	200.00	440.00	240.00	1.40	97.20
Yanadi Colony	7.90	340.00	120.00	500.00	980.00	1.00	53.40
Srinivas Nagar	7.80	432.00	132.00	220.00	420.00	1.60	72.90
Kamanur	7.80	320.00	132.00	588.00	2640.00	1.00	45.60
Radha Nagar	7.90	420.00	132.00	440.00	520.00	0.70	69.90
Nakkaladinne	8.00	420.00	140.00	700.00	632.00	1.60	68.00
Jamespeta	8.10	420.00	160.00	300.00	700.00	1.10	63.10
Chinnamurujupalli	7.70	500.00	156.00	580.00	700.00	1.30	83.50
Nagaiahpalli	7.90	200.00	80.00	294.00	1884.00	0.60	29.10
Kakirenipalli	7.70	312.00	104.00	430.00	380.00	1.50	50.50
Narasimhapuram	7.80	384.00	140.00	600.00	380.00	1.40	59.20
Upparapalli	7.60	304.00	104.00	770.00	332.00	1.70	48.60
Bojjavaripalli	7.70	320.00	104.00	510.00	996.00	0.80	52.40

Bankachinnaipalli	8.00	260.00	92.00	716.00	1480.00	1.00	40.80
Thallamapuram	8.00	592.00	196.00	1080.00	1160.00	0.90	96.20
Sagiliguddupalli	7.80	340.00	120.00	470.00	324.00	1.50	53.40
Mallavaripalli	8.10	840.00	380.00	592.00	300.00	1.00	111.70
Seethampalli	8.10	592.00	192.00	400.00	440.00	1.30	97.20
Kalluru	7.70	424.00	156.00	460.00	260.00	0.80	65.10
Nagganuripalli	7.50	252.00	100.00	580.00	600.00	1.30	36.90
Kothapalli	7.80	908.00	320.00	420.00	340.00	1.60	142.80
Lingapuram	8.20	392.00	132.00	220.00	1400.00	1.30	63.10
Kadarbad	8.20	320.00	108.00	420.00	620.00	1.10	51.50
Menapuram	7.80	312.00	104.00	650.00	1200.00	1.60	50.50
Kanapalli	7.80	700.00	480.00	400.00	692.00	1.10	53.40
Vivekananda nagar	8.10	696.00	220.00	436.00	432.00	1.60	115.60
Easwar Reddy nagar	7.80	220.00	88.00	360.00	2540.00	2.50	32.00
Penna Nagar	8.00	260.00	92.00	480.00	408.00	1.30	40.80
Somulavaripalli	7.90	284.00	80.00	310.00	5000.00	1.10	49.50
Settipalli	7.60	340.00	120.00	306.00	392.00	0.70	53.40
Bollavaram	7.50	232.00	76.00	320.00	532.00	0.80	37.90

Table 2: Summary statistic of hydrochemical data before log transformation

Variable	Min	Max	Mean	Median	SD	Skewness	Kurtosis
pH	7.4	8.2	7.8674	7.9	0.18609	-0.41294	2.8482
Total Hardness	200	1692	433.67	340	253.35	3.1767	15.446
Calcium	76	520	155.81	132	96.822	2.5121	8.9213
Magnesium	29.1	284.7	67.472	53.4	41.504	3.6024	18.724
Chloride	180	5000	866.98	580	892.86	2.8114	12.209
Fluoride	0.6	2.5	1.2558	1.3	0.38273	0.65584	4.412
Alkalinity	200	1080	455.86	430	171.07	1.2363	5.4526

Table 3: Summary statistic of hydrochemical data after log transformation

Variable	Min	Max	Mean	Median	SD	Skewness	Kutosis
pH	2.0015	2.1041	2.0625	2.0669	0.023782	-0.47249	2.92.5
Total Hardness	5.2983	7.4337	5.9703	5.8289	0.41637	1.2732	5.1121
Calcium	4.3307	6.2538	4.9309	4.8828	0.44305	1.4094	4.7874
Magnesium	3.3707	5.6514	4.1064	3.9778	0.41889	1.3431	5.8561
Chloride	5.193	8.5172	6.4469	6.363	0.74515	0.7656	3.1036
Fluoride	-0.51083	0.9162	0.18658	0.26236	0.29434	-0.26774	2.8575
Alkalinity	5.2983	6.9847	6.0587	6.0638	0.35933	0.0569	2.9849

Table 4: Correlation matrix

Correlation matrix (Pearson (n)):							
Variables	PH	TH	Ca	ALKALINITY	CL	F	Mg
PH	1	0.167	0.150	-0.024	0.088	0.010	0.162
TH	0.167	1	0.905	-0.014	0.029	0.119	0.970
Ca	0.150	0.905	1	-0.003	-0.027	0.058	0.775
ALKALINITY	-0.024	-0.014	-0.003	1	-0.051	0.026	-0.019
CL	0.088	0.029	-0.027	-0.051	1	0.025	0.058
F	0.010	0.119	0.058	0.026	0.025	1	0.144
Mg	0.162	0.970	0.775	-0.019	0.058	0.144	1

Values in bold are different from 0 with a significance level alpha=0.05

Table 5: Factor analysis

Maximum change in communality at each iteration:	
Iteration	Maximum change
40	0.0005
41	0.0005
42	0.0005
43	0.0005

44	0.0004
45	0.0004
46	0.0004
47	0.0004
48	0.0004
49	0.0004

Table 6: Eigen values

	F1	F2	F3	F4	F5
Eigenvalue	2.802	0.287	0.174	0.030	0.002
Variability (%)	40.029	4.097	2.487	0.424	0.034
Cumulative %	40.029	44.127	46.614	47.038	47.071

Table 7: Best interpolation method based on RMSE

GWQI	IDW	RBF	Kriging
pH	0.1953	0.1942	0.1856
TH	284.5	271.6	308.9
Ca	105.7	102.3	108.2
Mg	46.88	44.61	51.45
Cl	964	911.9	985.7
F	0.3839	0.3813	0.3851
Alkalinity	173.6	172.5	173.8

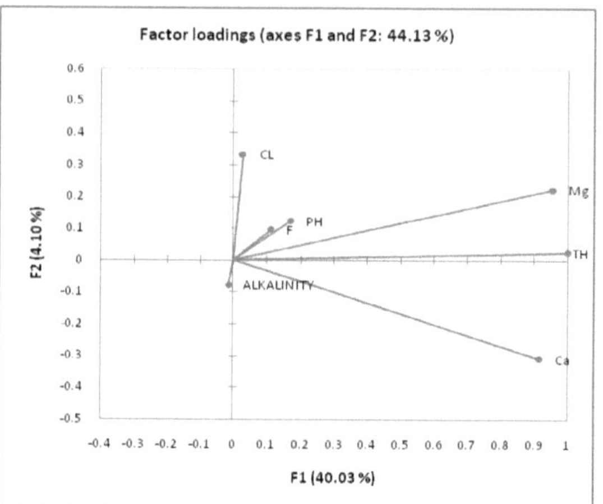

Figure 2: Factor loadings

Figure 3: Factor loadings after varimax rotation

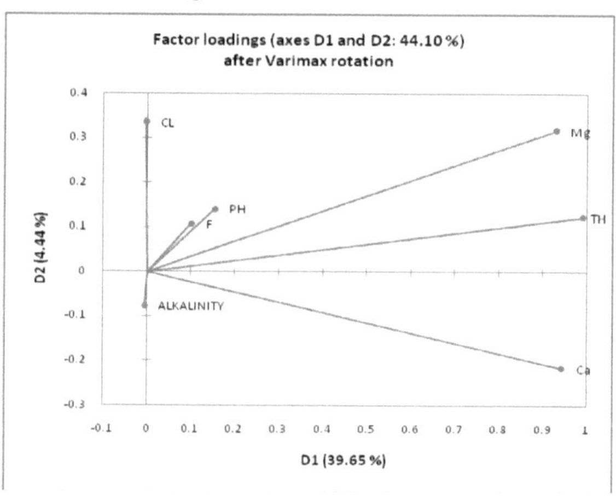

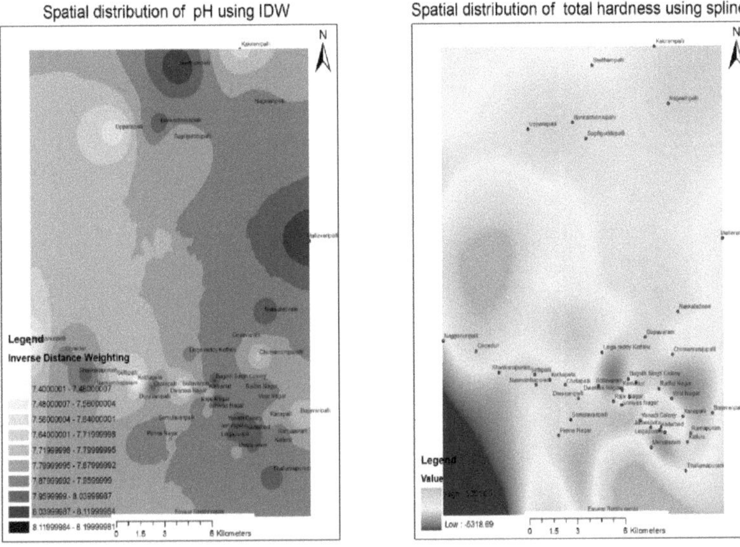

Figure 4: Spatial distributions of pH and total hardness

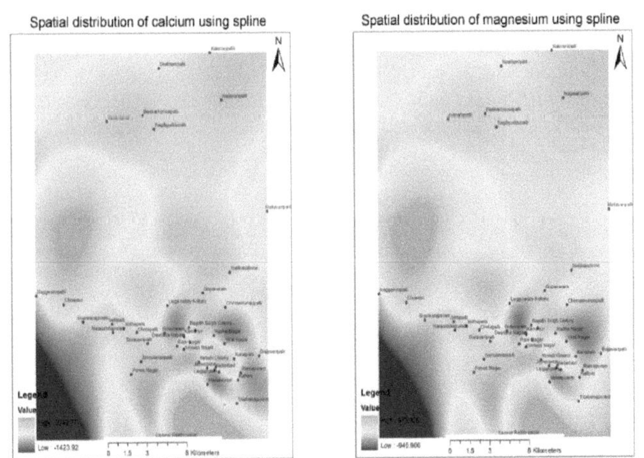

Figure 5: Spatial distributions of Calcium and Magnesium

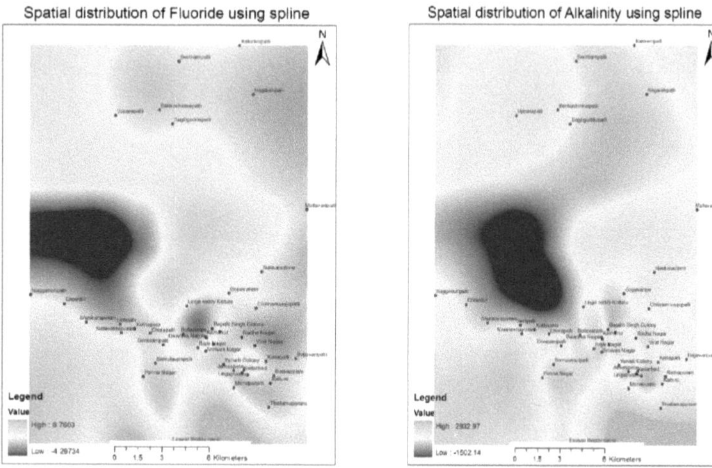

Figure 5: Spatial distributions of Fluoride and Alkalinity

5. Conclusion

The statistical treatment was applied to the hydro geochemical data via principal component analysis and factor analysis. Three interpolation techniques were critically analyzed and the best fit was selected after evaluating RMSE values. As per the magnitude of variations and prominence offered by the data, it is observed that except pH, all the other variables such as Total hardness, Calcium, Magnesium, Chloride, Fluoride and alkalinity accepts radial basis function method of interpolation and pH sticks on to the Inverse Distance Weighted method of interpolation. GIS maps were prepared for groundwater quality parameters.

Case study 3: GIS based groundwater quality mapping in central part of Proddatur town, Y.S.R Kadapa District, Andhra Pradesh, India

Abstract: The spatial variations in ground water quality in central part of Proddatur town which is located in Andhra Pradesh state, India, have been studied using geographic information systems. The study area lies in the Southwestern part of Y.S.R Kadapa District, Andhra Pradesh. The area is located in the survey of India toposheets No 57 J/6 lying between east $78^035'$ - $78^055'$ longitudes and $14^044'$ - $14^073'$ North latitudes. For this study, water samples were collected from 43 of the open wells and bore wells which represent the entire municipality area. The samples were systematically analyzed for physico-chemical parameters such as pH, TH, Ca, Mg, Cl, F and alkalinity. It is observed that ground water quality of the study area is deteriorating beyond the limits and hence can affect human health adversely if not properly mitigated. The maps were prepared using Spatial Analyst and Geostatistical Analyst extensions of ARC-GIS 9.3. The results obtained in present study with spatial database established in GIS will be used to monitor effectively and periodically assess the quality and vital parameters of ground water and also to draw the attention of civic authorities for suitable action.

Keywords: Geostatistical Analyst, Groundwater quality, Descriptive statistics, Physico-chemical analysis, spatial analysis, Interpolation, Hydrochemical Maps, Proddatur.

1. Introduction

Ground water is an essential renewable resource which we rely on since centuries without estimating its fate in terms of quality and quantity. The ground water is considered as pollution free and cleaner than surface water due to its hideous nature underneath the surface. Industries, residential, municipal, agricultural and commercial activities adversely affect the quality of ground water. Cases' reflecting huge demand for water, land, health, transport, education and housing is not uncommon these days due to rapid urbanization. The sustenance of life depends on the availability of fresh water and its protection for posterity. Over exploitation of ground water is of primary concern due to its pronounced an irrational use in the recent decades. With this present pace of ground water deterioration in terms of quality and quality, there would be a worldwide alarming effect within less than a century; give or take. Ground water is a valuable natural resource that is essential for human health, socio-economic development and functioning of ecosystems (Zekster, 2000; Humphreys, 2009; Steube et al., 2009).80% of the population in Indian sub-continent suffer from diseases due to poor drinking water quality and unhygienic conditions.(Olajire and Imeokparis, 2001; Prasad, 1984).Its proved mandatory to overdepend on ground water due to ever- increasing demands from the key sectors such as agriculture, domestic and industries. Geographical information systems have proved their dominance in present day scenarios with the ground water quality issues in every corner of the world.

Considering serious aspects of ground water quality issues in the urban areas it is essential to use advanced tool such as GIS. The present study was undertaken to map the quality of ground water in Proddatur town, Andhra Pradesh, India. The literature survey indicates that earlier researchers have made studies on ground water issues using data obtained from open wells and bore wells in the town. It is further observed that there is less considerable work undertaken to use GIS on mapping and geostatistical analysis of the ground water quality in the study area. With all due respect to the earlier researchers in the study area, it has been sensed that it is quintessential to focus on spatial variations of certain physico-parameters via GIS platform. The main objective of this study is to assess the ground water quality based on physico-chemical data from 43 locations in Proddatur town. The important objectives of this assessment include providing an overview of ground water quality in current situation and determining spatial distribution of key ground water quality parameters such as pH, TH, Ca^{2+}, Mg^{2+}, Cl^-, F^- and alkalinity.

2. Aim and Objectives

The main aim is to identify groundwater quality in central part of Proddatur town, YSR Kadapa District through,

- Descriptive statistical analysis
- Spatial Interpolation
- Geostatistical analysis.

3. Methodology

In the present study, water samples were collected from forty three wells in the study area. The study has been carried out using topographic sheets, Arc GIS 9.3, Garmin GPSMAP76 and relevant fieldwork. Base map and drainage map were prepared using toposheets n order to understand the nature of the location. The GPS coordinates were exported on to the GIS platform for further analysis. The water samples were collected during February 2012.The analysis of ground water samples were performed according to the procedure of APHA (1995), and necessary steps were followed to prevent contamination. The key parameters determined were: pH, Total hardness (TH), Calcium (Ca), Magnesium (Mg), Chloride (Cl), Fluoride (F) and alkalinity. The spatial variation of the ground water quality parameters were studied using spatial analyst and Geostatistical analyst extensions modules of ARC-GIS 9.3. The well locations were established in the GIS environment and the results of the every parameter under study were added to the concerned well.

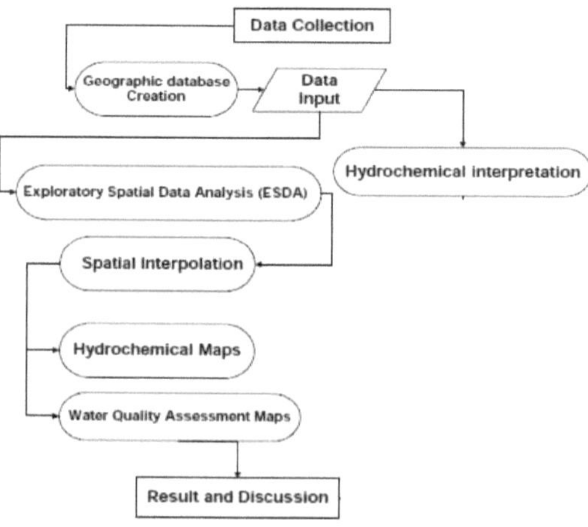

Figure 1: Main implementation steps

3.1 Study Area

The study area (Fig. 2) in the southwestern part of Y.S.R Kadapa District of Andhra Pradesh is in between east $78^{0}33'$ - $78^{0}55'$ longitudes and $14^{0}44'$ - $14^{0}73'$ North latitudes and falls in the Survey of India Toposheet no. 57 J/6.It has an average elevation of 155 meters. It is on the banks of river Penna.It is considered to be second Bombay for it is a commercial hub gold and cotton businesses.

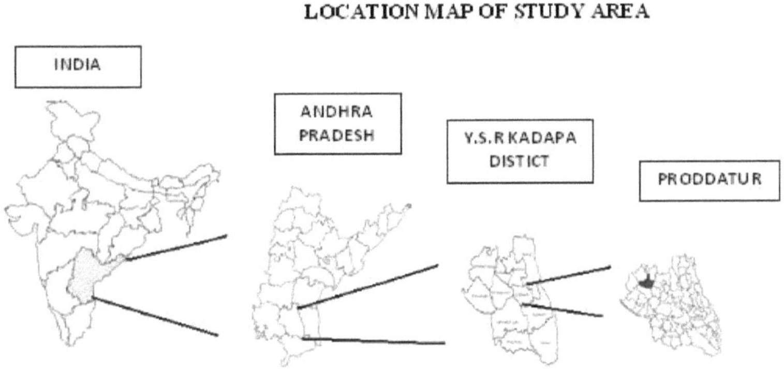

LOCATION MAP OF STUDY AREA

Figure 2: Study Area

4. Results and Discussion

The results of the physico-chemical parameters were analyzed and represented in table 1.The following statistical parameters which include minimum, maximum, mean, median, standard deviation, skewness and kurtosis were studied. The analysis was carried out keeping view of the importance of the ground water quality which determines its suitability for agricultural, industrial purposes. The pH values of the ground water ranges from 7.4 to 8.2 with an average of 7.86 during February 2012. Pronounced alkalinity was also observed from the data which ranges from 200 to 1080 mg/l. The total hardness values ranges from 200 to 1692 mg/l during February 2012 with an average of 433.67.The chloride value ranged between 180 to 5000 mg/l with an average of 866.98 during February 2012.The fluoride values ranged between 0.6 to 2.5 mg/l with an average of 1.25 during the same period. The magnesium and calcium values were 67.42 and 155.81 mg/l at an average respectively. The analytic results of physico-chemical parameters were compared with World Health Organization standards (WHO 1983) and Indian Standard (ISI 1983) for drinking and public health purposes. (Table 2) It is observed that the values obtained keenly reflect that the key parameters were well above the recommended limits.

The GIS based spatial variation behavior in the ground water quality is studied using spatial analyst and geostatistical analyst modules of Arc GIS 9.3 after establishing geodatabase. The analysis was done using interpolation method such as inverse distance weighted (IDW).This method is an algorithm for spatially interpolating values which are estimated between measurements. In IDW, every value is a weighted average of surrounding points. Weights are computed by taking the inverse of the distance from observations location to the location of the point being estimated (Burrough and MC Donnell 1998).IDW determines cell values by using linearly weighted combination of a set of points. The Geostatistical analyst was used to create continuous surface or a map from the measured sample points which are stored in a point-feature layer. This module will provide a comprehensive set of tools which can be used to visualize, analyze and determine spatial phenomena. The surface fitting involves exploratory spatial data analysis, structural analysis, surface prediction and assessment of results.

The spatial pattern of the physico chemical parameters was studied in February 2012 in the study area. The spatial analysis of pH during the study period shows the variation in spatial highs and lows. High pH exists in eastern and northern part and low pH is observed in the western part.   The generated surface of the total hardness was classified according to Bouwer,1978 as illustrated in output, where the hard water exist in smaller area in north part of the study area and less in south. The magnesium and calcium concentrations are higher in eastern and western parts of the study area. The chloride concentration is high in south west and north east part of the study area. Pronounced fluoride concentration exists in North West, south and eastern part of the study area.

Villages	PH	TH	Ca	Alkalnity	Cl	F	Mg
Chowdur	7.90	1692.00	520.00	310.00	2420.00	1.50	284.70
Kothapeta	7.40	320.00	108.00	200.00	280.00	1.10	51.50
Shankarapuram	8.00	352.00	132.00	240.00	580.00	1.60	53.40
Chotapalli	8.00	420.00	140.00	380.00	1180.00	1.00	68.00
Dorasanipalli	7.80	320.00	124.00	680.00	1040.00	1.50	47.60
Ramapuram	8.00	308.00	104.00	400.00	420.00	1.00	49.50
Gopavaram	7.90	300.00	104.00	420.00	580.00	1.60	47.60
BagathSingh Colony	8.10	400.00	160.00	320.00	320.00	1.10	58.30
Virat Nagar	7.90	300.00	100.00	360.00	180.00	0.90	48.60
Linga reddy Kottalu	7.90	592.00	196.00	520.00	236.00	1.50	96.20
HousingBoard Colony	7.90	400.00	140.00	440.00	580.00	1.60	63.10
Rajiv Nagar	8.00	308.00	112.00	340.00	540.00	1.40	47.60
Dwaraka Nagar	7.70	600.00	200.00	440.00	240.00	1.40	97.20
Yanadi Colony	7.90	340.00	120.00	500.00	980.00	1.00	53.40
Srinivas Nagar	7.80	432.00	132.00	220.00	420.00	1.60	72.90
Kamanur	7.80	320.00	132.00	588.00	2640.00	1.00	45.60

Radha Nagar	7.90	420.00	132.00	440.00	520.00	0.70	69.90
Nakkaladinne	8.00	420.00	140.00	700.00	632.00	1.60	68.00
Jamespeta	8.10	420.00	160.00	300.00	700.00	1.10	63.10
Chinnamurujupalli	7.70	500.00	156.00	580.00	700.00	1.30	83.50
Nagaiahpalli	7.90	200.00	80.00	294.00	1884.00	0.60	29.10
Kakirenipalli	7.70	312.00	104.00	430.00	380.00	1.50	50.50
Narasimhapuram	7.80	384.00	140.00	600.00	380.00	1.40	59.20
Upparapalli	7.60	304.00	104.00	770.00	332.00	1.70	48.60
Bojjavaripalli	7.70	320.00	104.00	510.00	996.00	0.80	52.40
Bankachinnaipalli	8.00	260.00	92.00	716.00	1480.00	1.00	40.80
Thallamapuram	8.00	592.00	196.00	1080.00	1160.00	0.90	96.20
Sagiliguddupalli	7.80	340.00	120.00	470.00	324.00	1.50	53.40
Mallavaripalli	8.10	840.00	380.00	592.00	300.00	1.00	111.70
Seethampalli	8.10	592.00	192.00	400.00	440.00	1.30	97.20
Kalluru	7.70	424.00	156.00	460.00	260.00	0.80	65.10
Nagganuripalli	7.50	252.00	100.00	580.00	600.00	1.30	36.90
Kothapalli	7.80	908.00	320.00	420.00	340.00	1.60	142.80
Lingapuram	8.20	392.00	132.00	220.00	1400.00	1.30	63.10
Kadarbad	8.20	320.00	108.00	420.00	620.00	1.10	51.50
Menapuram	7.80	312.00	104.00	650.00	1200.00	1.60	50.50
Kanapalli	7.80	700.00	480.00	400.00	692.00	1.10	53.40
Vivekananda nagar	8.10	696.00	220.00	436.00	432.00	1.60	115.60
Easwar Reddy nagar	7.80	220.00	88.00	360.00	2540.00	2.50	32.00
Penna Nagar	8.00	260.00	92.00	480.00	408.00	1.30	40.80
Somulavaripalli	7.90	284.00	80.00	310.00	5000.00	1.10	49.50
Settipalli	7.60	340.00	120.00	306.00	392.00	0.70	53.40

Bollavaram	7.50	232.00	76.00	320.00	532.00	0.80	37.90
Min	7.4	200	76	200	180	0.6	29.1
Max	8.2	1692	520	1080	5000	2.5	284.7
Mean	7.86	433.6	155.81	455.8	866.98	1.25	67.472
Median	7.9	340	132	430	580	1.3	53.4
Standard deviation	0.186	253.3	96.82	171.07	892.86	0.382	41.50
Skewness	-0.412	3.176	2.512	1.236	2.811	0.655	3.602
Kurtosis	2.848	15.446	8.921	5.456	12.209	4.412	18.72

Table 1: Result of chemical and statistical analysis of physico-chemical parameters of ground water

Parameters	WHO International standard (1983)	Indian Standards for drinking water specification (IS: 10500, 1992)	Mean value of parameters under analysis
pH	7-8.5	6.5-8.5	7.86
TH (mg/l)	100	300	433.6
Cl^- (mg/l)	200	250	866.98
F^- (mg/l)	1.5	1.5	1.25
Mg^{2+} (mg/l)	50	30	67.42
Ca^{2+} (mg/l)	75	75	155.81
Alkalinity(mg/l)	30	200	455.8

Table 2: Comparison of ground water quality of the study area with WHO and Indian standards.

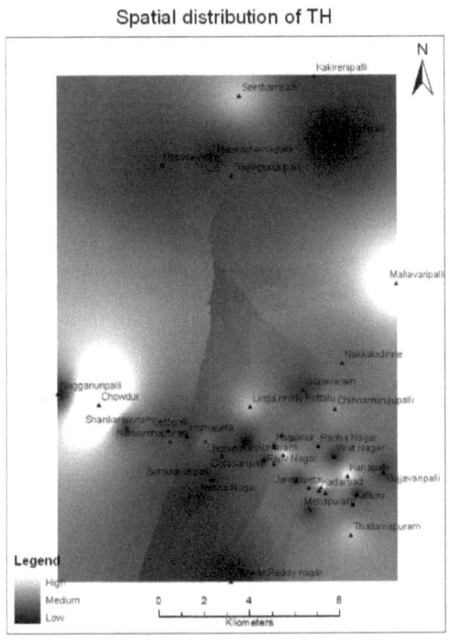

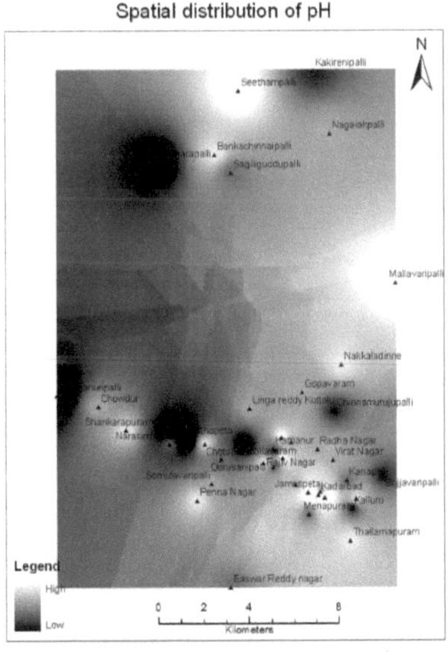

Figure 3: Maps showing spatial variation of pH and TH in the study area

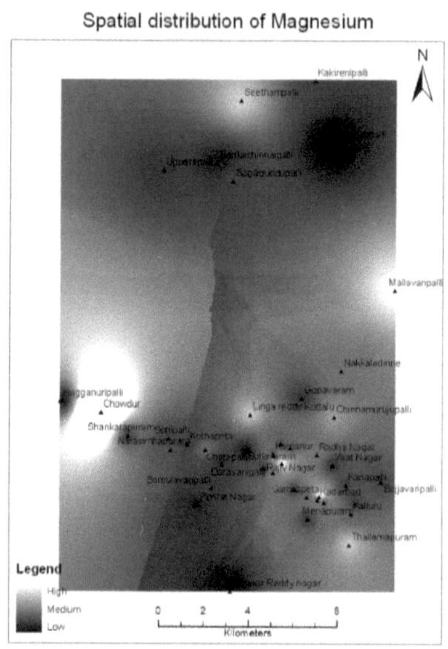

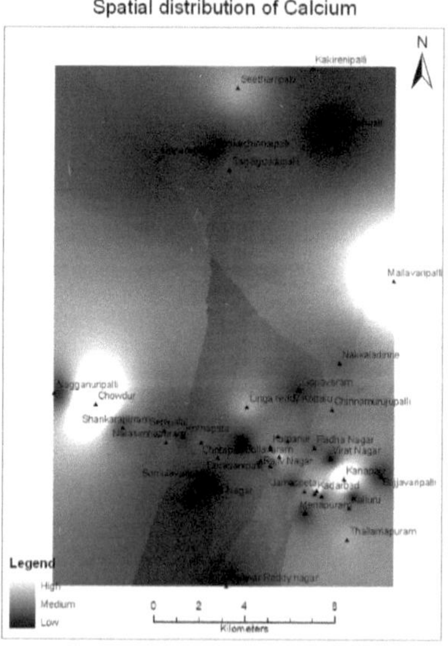

Figure 4: Maps showing spatial variation of Ca^{2+} and Mg^{2+} in the study area

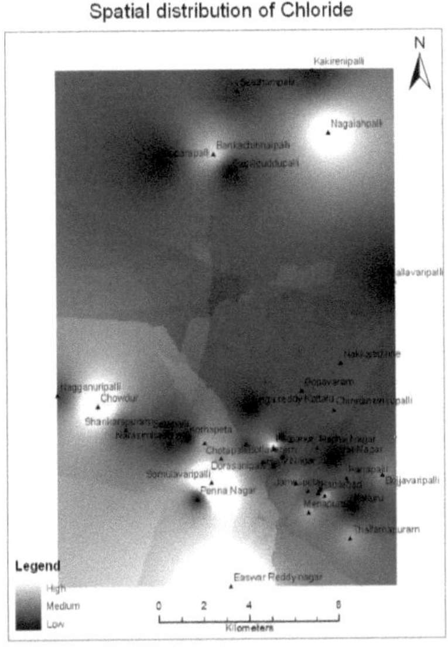

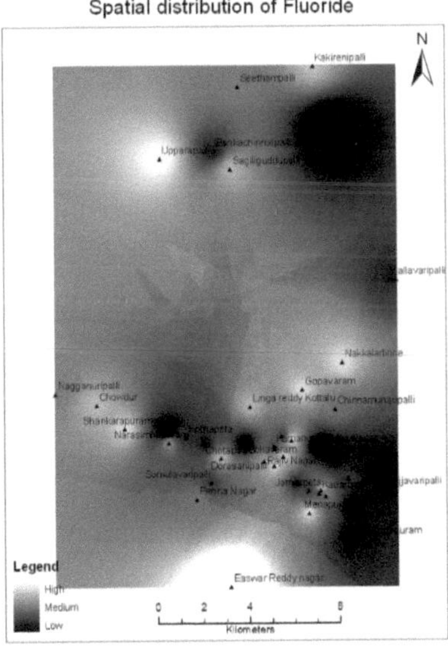

Figure 5: Maps showing spatial variation of Cl⁻ and F⁻ in the study area

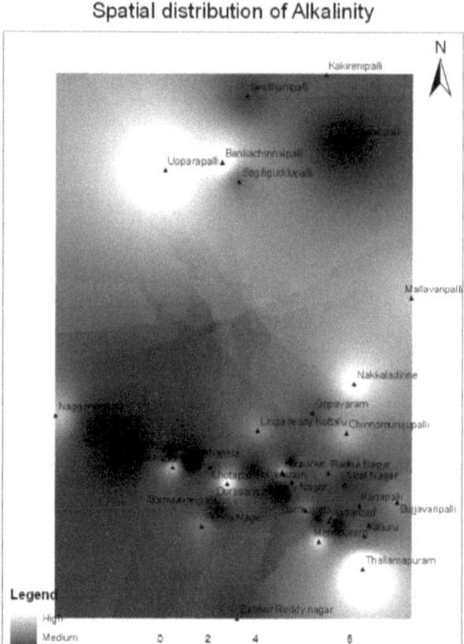

Figure 6: Maps showing spatial variation of alkalinity in the study area

5. Conclusion

The main objective of the present study was to understand and analyze the quality of ground water in central part of Proddatur town, Y.S.R. Kadapa district, Andhra Pradesh, India. The present work was designed to pictorially represent the spatial patterns using GIS of the ground water quality of the study area. The physico-chemical characteristics of the water samples reveal that the waters in this region are alkaline. All parameters analyzed except pH and fluoride are above the desirable limits of WHO and Indian Standards for drinking water. Hence the groundwater in the study area is not suitable for drinking. Rapid urbanization has lead to deterioration of the water quality in the study area. There is a need of sustainable measures to be taken by governing authorities to mitigate continuous deterioration of ground water quality.

Acknowledgement

Authors thank Dr R.V.Jayanth Kasyap, Assistant Professor, Department of English, Yogi Vemana university, India for providing support on linguistic appropriateness and grammatical corrections whilst manuscript preparation. We extend our gratitude to Dr B.Muralidhara Reddy, Research Scholar, Department of Geology and Geoinformatics for his support in processing the manuscript.

6. References

1. Gyananath G, Islam S.R: Shewdikar S.V, (2001), Assessment of Environmental Parameter on ground water quality, Indian journal of environmental protection, 289-294.

2. APHA. (1995), Standard methods for the examination of water and wastewater, 19th edn. American public Health Association, Washington.

3. Das Gupta, M, Purohiy, K.M,;Jayita Datta, (2001), Assessment of drinking water quality of river Bahmani, Journal of Environmental and pollution, 8,285-291.

4. Ducci D. (1999), GIS techniques for mapping groundwater contamination risk. Natural Hazards, 20, pp.279-294.

5. ESRI. (2002), Using Arc GIS spatial analyst. ESRI Press, Redlands.

6. Ahn H. (1999), Assessment of groundwater contamination using geographic information systems, Environmental Geochemical and Health, 21, pp.273-289.

7. Indian Standards Institution (1983), Indian standard for drinking water IS 10500.

8. Babiker IS, Mohammed AM, Hiyama T, (2007), Assessing groundwater quality using GIS, Water resource management.,21(4), 699-715.

9. Nas B, Berktay A. (2010), Groundwater quality mapping in urban groundwater using GIS Environmental Monitoring and Assessment., 160(1-4), pp.215-227.

10. Srinivasa Gowd S (2005), Assessment of groundwater quality for drinking and irrigation purposes: a case study of Peddavanka watershed, Anantapur District, Andhra Pradesh, Indian Journal of Environmental Geology 48, 702-712.

11. Srivastava PK, Bhattacharya AK. (2000), Delineation of groundwater potential zones in a hard terrain of Bargarh district, Orissa using IRS Data, Journal of Indian Society of Remote Sensing, 28(2), 129-140.

12. Subramani T, Elango L, Damodarasamy S.R. (2005), Groundwater quality and its suitability for drinking and agricultural use in Chithar River Basin, Tamil Nadu, Indian Journal of Environmental Geology, 47, 1099-1110.

13. Sujatha D, Rajeswara Reddy B, (2003), Quality characterization of groundwater in the south-eastern part of the Ranga Reddy district, Andhra Pradesh, Indian Journal of Enviromental Geology, 44, pp.579-586.

14. Todd DK. (2007), Groundwater Hydrology, Wiley- India Edition.

15. Shankar.K, Aravindan.S, Rajendran.S. (2010), GIS based Groundwater Quality Mapping in Paravanar River Sub- Basin, Tamil Nadu, India, International Journal of Geomatics and Geosciences, 1(3), pp. 282-296.

16. World Health Organization (1983), International Standards for Drinking Water, Geneva.